A Manual for Designers and Builders

STRAW BALE DETAILS

Chris Magwood and Chris Walker

Cataloguing in Publication Data:
A catalogue record for this publication is available from the National Library of Canada.

Cover design and layout by e_e_e@sympatico.ca
Cover art © 2001 by Emese Ungar.

Printed in Canada.

New Society Publishers acknowledges the support of the Government of Canada through the Book Publishing Industry Development Program (BPIDP) for our publishing activities, and the assistance of the Province of British Columbia through the British Columbia Arts Council.

BRITISH
COLUMBIA
ARTS COUNCIL
Supported by the Province of British Columbia

Spiral-bound ISBN: 0-86571-476-2

Inquiries regarding requests to reprint all or part of *Straw Bale Details* should be addressed to New Society Publishers at the address below.

To order directly from the publishers, please add $4.50 shipping to the price of the first copy, and $1.00 for each additional copy (plus GST in Canada). Send cheque or money order to:

New Society Publishers
P.O. Box 189, Gabriola Island, BC V0R 1X0, Canada

New Society Publishers aims to publish books for fundamental social change through nonviolent action. We focus especially on sustainable living, progressive leadership, and educational and parenting resources. Our full list of books can be browsed, and purchased securely, on the worldwide web.

NEW SOCIETY PUBLISHERS
www.newsociety.com

TABLE OF CONTENTS

Since the early 1990s, the popularity of straw bale building has grown substantially. This growth has been different from the growth of other building systems in that it has not been driven by patented or aggressively marketed technologies. At no point in the straw bale revival have significant amounts of capital been invested by businesses or governments. Straw bale construction is truly a grassroots movement.

This has allowed for a great deal of experimentation, trial and error, and hands-on learning in the field. The ways of building with straw bales are as numerous as the number of straw bale buildings currently standing. However, enough buildings have now been successfully completed that bale designers and builders no longer need to keep reinventing the wheel with each new structure. Without stifling the creative options available to designers and builders, certain key details have proven themselves to be effective, simple and cost-effective. Conversely, the exclusion of these important details has lead to some buildings experiencing everything from minor glitches to complete failure.

Much has been made of the learner-friendly nature of straw bale building. While it is true that the physical stacking of bales and, to a lesser degree, applying plaster to bale walls, are easily comprehended skills, straw bale building is not exempt from the many important detailing considerations drawn from centuries of collective construction experience. Poorly built bale buildings are no better than poorly built buildings of any other material. Inexperienced owner-builders are very capable of making excellent, long lasting bale buildings, as long as they are detailed well.

The modern application of straw bale building technology is still in a phase of rapid development. The drawings and notes included in this booklet represent concepts derived from the existing base of scientific study and specific details which have been successfully applied in the field. New approaches will continue to be field-tested, and the information included in this booklet will be updated accordingly. Input from designers and builders is always welcome.

No attempt has been made in this booklet to comment on design, rather, its purpose is to expertly detail any design so it is well built and will stand the test of time. It is not difficult to design and construct a well detailed bale building, but the details are not always obvious at the paper stage, and it often becomes time consuming and bothersome to fix things once they've been built according to plans without proper detailing. At best, this adds time and cost to the project, at worst, it means these details are ignored, often with negative consequences for the building.

This manual is not meant to be the final word on detailing for straw bale building. Hopefully it will see many useful additions and revisions in the years to come, as this promising technology is put to use by more and more builders. Working with the available test results and a knowledge of sound building practices, there are countless variations to the details outlined here which can be successfully implemented. The details in this book have all been put to use in the field many times, and as such represent good, achievable and affordable options.

The environmental benefits of building with straw bales are well documented, and are a key motivating factor for many owners and designers who choose bale walls. Obviously, the selection of other materials for such buildings should also be made with environmental criteria in mind. Many environmentally-friendly building materials are relatively new and are not always obvious or widely available. I encourage you to source and apply these materials, and to share the results with other designers and builders. We all look forward to the day when the built environment does not require such devastation of the natural environment, and it is up to all of us to go out of our way to create examples of such buildings.

Contact   We encourage you to share your results, reactions and developments to these details with us and with the natural building world at large. Acting collectively, we can speed the progress of more sustainable building from the alternative fringe to mainstream practice, where these ideas belong.

Chris Magwood
R.R. 3
Madoc ON K0K 2K0
cmagwood@kos.net

Chris Walker
176 Borden Street
Toronto, ON M5S 2N3
walkino@canoemail.com

Acknowledgements   To Peter Mack, Tina Therrien, Kris Dick, the people whose homes we've designed, and/or built, Catherine Wanek, the Black Range Lodge, *The Last Straw Journal*, Bruce King, Jeff Rupert, Matts Myhrman, Judy Knox, Joyce Coppinger, David Eisenberg, to all the workshop participants, and our family and friends, our deepest thanks for all your efforts and shared thoughts.

KEY SCIENTIFIC TEST RESULTS

A significant number of testing programmes have been carried out on many issues crucial to understanding the strengths and limitations of straw bale walls. The majority of these tests have been done on a small scale, with minimal funding, and we would all benefit from follow-up tests. Between the test results garnered to date and the examples of many historic and well-aged bale buildings, we know a great deal about how these walls work, and how and when they fail.

Included below are excerpts from some of the more informative tests. These are included to sketch an outline of the acceptable parameters of straw bale walls, but they are no replacement for reading the entire test documents themselves, which we encourage you to do, using the bibliography at the end of this booklet.

Structural Testing

Structural testing carried out by a number of different laboratories in several countries have all determined that straw bales function as structural fibre insulation in a stress-skin panel, created by the application of a plaster finish directly to the bales on the inside and outside of the walls. Understanding the principle of the stress-skin panel is key to designing and building a sound bale structure, and applying the lessons learned in stressing the walls to failure is the aim of all the details in this booklet. The most significant structural testing results garnered to date are given below.

Compression / Vertical Load Test Results

Compression testing of plastered bale walls at the University of Colorado at Boulder in 1999 showed the following average results for 8-foot-high and 12-foot-long bale walls without wire mesh reinforcement in the plaster skins:

three string bales	38,867 pounds ultimate strength
	3,239 pounds per lineal foot
two string bales	73,877 pounds ultimate strength
	6,156 pounds per lineal foot

In summarising the results of the test, the authors noted that several factors *resulted in a stucco quality lower than will generally be encountered in the field, rendering these results conservative.* These values can be used as a conservative guideline to determine the load-bearing capacity of a straw bale wall system without any internal or external framing system.

Racking Load Test Results

The Building Research Centre at the University of New South Wales conducted tests on two-string bale walls according to ASTM E72 standards, stating that *in both the racking tests carried out, the 10 kN (2,248 lb/ft) horizontal load at the top corner produced small deflections of slightly more than 2mm. The material's performance under racking load would be considered acceptable.*

Transverse / Wind Load Test Result

The University of New South Wales used the vertical chamber method specified in ASTM E72 to test the walls to a maximum wind speed equivalent of 58 m/s (130 mph) stating that *the maximum static air pressure of 2.5 kPa (0.4 lb / in²) that was applied represents a significant wind of over 60m/s. Both walls tested showed small deflections of around 7 mm (¼"). The walls would be considered acceptable for structural behaviour under wind load.*

These results show that load-bearing straw bale walls offer enough strength in every direction to meet equivalency requirements when compared to traditional stud framing. The most important lesson learned in these tests is that the majority of any loads applied to the bale wall are handled by the plaster skin. Therefore, it is imperative that both the interior and exterior plaster skins be positioned in such a way that roof loads are received into the plaster skins, and that the skins transfer the load to the foundation. All top plates and foundation details in this booklet attempt to do just this, as should any variations adapted by designers or builders.

Load-bearing straw walls have been successfully built to heights of 3.35 m (11') and uninterrupted lengths of 23 m (75'). They have been used to support trusses with a clear span of 12.8 m (42'). These figures suggest that load-bearing walls can function adequately in most residential construction scenarios.

Key to the popularity of straw bale building is the insulative qualities of the wall system. While test result figures for the R-value of a bale wall vary wildly, experience has shown that, regardless of any given figures, a well built bale wall exceeds the common standards for insulative value, placing bale walls in the *super-insulated* category.

Several tests have been undertaken to establish the insulative value of a straw bale wall. The results have varied, but the testing performed for the Canadian Society of Agricultural Engineering was carried out on an existing bale building, reflecting real world values. The results stated that the *R-value, in FPS units, varied from 30 to 40. The R-value of the straw bale walls is in the range of super efficient homes. More testing needs to be done to acquire more definitive values.*

Figures in other laboratory tests have ranged from R-26 to R-52. More important than the actual R-value is the need to ensure that bale walls are free from gross air leakage on the inside and outside. All the details in this booklet attempt to minimise or eliminate potential air leakage into the wall.

SHB Agra, Inc., conducted a small scale fire test according to ASTM E-119 standards which stated that *the plastered bale panel was tested for over two hours and withstood temperatures that reached 1942°F. The temperature rise on the unheated side of the test panel, after 2 hours, averaged less than 10°F, with the highest being 21°F. It is clear from these results that fire resistivity is a potential benefit, rather than a problem for straw bale wall systems.*

While fire is often the foremost concern for those new to bale building, the walls actually represent a remarkably fire-resistant system, with burn times that meet or exceed commercial fire ratings.

The most comprehensive tests of moisture levels and resultant straw condition were completed by Rob Jolly for the Canada Mortgage and Housing Corporation. The test included the monitoring of nine different buildings over a two year period. He concluded that *straw bale walls do not exhibit any unique propensity for moisture retention. It is clear that straw bale walls can function, without incorporating an interior vapour barrier, in northern climates that receive mild to moderate amounts of precipitation. In comparison to standard frame construction, straw bale walls generally incorporate higher perm (more breathable) interior and exterior protective layers. Combined with the hygroscopic nature of straw, these factors allow for a highly dynamic wall system. Within limits, a straw bale wall has the capacity to ad/absorb moisture, and diffuse this moisture to either the exterior or interior of a structure. However, this capacity should not be used as an excuse for inappropriate designs or applications.*

01 **A primary design consideration must continue to be the protection of the straw from exterior wetting.**

02 **When designs are inappropriate, straw bale walls can fail, even in low precipitation climates.**

03 **Successful designs, in terms of moisture control, monitored and observed during this study incorporated combinations of the following diverse strategies:**

   a. The use of verandahs and oversize gables and overhangs for direct precipitation protection

   b. Standard sized gables and overhangs in conjunction with a combination of other factors:

   - appropriate protection afforded by topography and vegetation

   - minimal wind driven precipitation

   - infrequent exterior wetting accompanied by prolonged drying cycles

   c. The use of vapour permeable sheet moisture barriers in conditions where enough exterior moisture protection has not been provided by other means

   d. Cement based parging used on exterior and interior

   e. Sheathing the exterior bale surface with plywood, then using building paper and cement based stucco

   f. Building paper, used as a vapour retarder on the interior prior to parging

   g. Cement based exterior parging, earthen plaster interior

   h. Bales elevated well above grade to eliminate backsplash

04 **Designs which produced borderline or unacceptable moisture readings included two or more of the following:**

   a. Minimal or absent overhangs

   b. No capillary break between foundation parging and above grade stucco

   c. Structures subject to extreme interior wetting without drainage

   d. Below-grade bales

   e. Inadequate backsplash protection

   f. Northern exposures

05 **Walls with southern exposures were generally much drier than other exposures and were able to handle significantly more exterior wetting.**

06 **Interior humidity control seemed to have little effect on exterior bale moisture content.**
   NOTE: This is not to say that a lack of interior venting is a recommended building practice.

07 **Extreme diurnal variances in RH (with spikes as high as 98%) do not seem to be indicative of straw degradation. Moisture content values correlated most closely with the minimum daily RH value. Prolonged high RH values (over 85%) generally indicates a problem.**

It was recommended in the previously submitted report that hydrophobic coatings (ie. acrylic-based stuccoes) at the exterior stucco surface may be an appropriate means to prevent wetting. None of the houses in this study incorporated this strategy. Although many of these products are highly vapour permeable, they also act as a capillary break at the exterior stucco surface. Significant leaks, due to poor detailing, will dry only through vapour diffusion and not through a combination of diffusion and capillary action. Although sheet moisture barriers have been used successfully in two houses in this study, a similar caution should apply.

Conclusion

The details included in this booklet attempt to observe all of the conclusions drawn from this testing. However, no details are given for the use of sheet moisture barriers. All of the structural testing noted above relied on the straw-plaster bond uninterrupted by sheet barriers to create the stress skin panel condition which produced the structurally sound results. By separating the plaster skin from the bales, sheet barriers can potentially compromise the strength of the wall system, and for that reason they have not been included.

KEY BUILDING CODE PROVISIONS

The State of California passed a bill adding provisions for straw bale construction to its building code in 1995. Other American jurisdictions to include provisions for straw bale construction in their building codes include New Mexico, Colorado, Texas, Arizona and Nevada. The codes in these jurisdictions are very similar to one another.

Selection of Bales

18944.35

b. Bales used within a continuous wall shall be of consistent height and width to ensure even distribution of loads within wall systems.

d. The moisture content of bales, at time of installation, shall not exceed 20 percent of the total weight of the bale. Moisture content of the bales shall be determined through the use of a suitable moisture meter, designed for use with baled straw or hay, equipped with a probe of sufficient length to reach the centre of the bale, and used to determine the average moisture content of five bales randomly selected from the bales to be used.

e. Bales in load-bearing walls shall have a minimum calculated dry density of 7.0 pounds per cubic foot. The calculated dry density shall be determined after reducing the actual bale weight by the weight of the moisture content.

f. Where custom-made partial bales are used, they shall be of the same density, same string or wire tension, and, where possible, use the same number of ties as the standard size bales.

g. Bales of various types of straw, including wheat, rice, rye, barley, oats, and similar plants, as determined by the building official, shall be acceptable if they meet the minimum requirements of this chapter for density, shape, moisture content and ties.

Orientation of Bales

18944.33

d. Laid flat refers to stacking bales so that the sides with the largest cross-sectional area are horizontal and the longest dimension of this area is parallel with the wall plane.

e. Laid on edge refers to stacking bales so that the sides with the largest cross-sectional area are vertical and the longest dimension of this area is horizontal and parallel with the wall plane.

Conclusion

While the existing code provisions for bale building are useful in setting an important precedent for the use of load-bearing straw walls they are also, understandably, quite conservative and, in places, already outdated. For this reason, I would not suggest that designers and builders operating outside jurisdictions covered by these codes feel obliged to follow them strictly.

BALE BASICS

Bale structures can utilise the stacked bale walls as the sole structural element, referred to as load-bearing or *Nebraska Style*, or they can be used in conjunction with a structural framework of lumber, steel or concrete. Details for both systems are included in this booklet. However, due to the extra materials and complexity of post and beam frame systems and to the numerous and well-documented ways in which frames can be built, details for post and beam buildings concentrate on the interface between frame and bales, not on actual frame construction.

Bales should be stacked in such a manner that a running bond is maintained, with the joints in each successive course in bales staggered by a minimum of 8". Full length bales should be used at all corners. For load-bearing designs, no door or window openings should be placed within a full bale length of any corner.

Internal pinning or spiking of bales, while common practice in early bale structures, is not a requirement. Pins or spikes can be used to aid in the stacking process, but are not a structural necessity. Most bale builders no longer use any internal pinning. Two preferable alternatives to internal pinning are, external pinning, whether permanent or temporary, and external bracing. Bale dimensions must be taken into account to determine the overall height of walls, especially in post and beam structures. Bale dimensions should also be used, whenever possible, to determine appropriate dimensions for rough door and window openings.

Roof overhangs for straw bale buildings should typically be 18"–24". Bales and/or the finished lower edge of the plaster finish must be a minimum of 8"–12" above grade. Eavestroughing or a suitable alternative must be provided to prevent excessive wetting of the walls.

Dimensions used for bale walls should correspond to actual measurements of bales. The most critical measurement is the width of the bale, which will determine the thickness of the wall. Height of the bale is more important for post and beam designs, where post heights should correspond to bale heights. Both width and height are fixed dimensions and will not vary much between bales from the same source. It is best to assume extra width. Length of bales is least critical, and most variable. Where actual bale dimensions are not known when designing, compensate with extra width. Two-string bales are often assumed to be 18" wide and 14" tall. More realistic dimensions are 19" wide and 14" tall, and should be the minimum used in drawings where actual dimensions are not known.

Plasters and Finishes

The plaster skins applied to straw bale walls provide the walls with their structural capacity, but they also determine the visual finish. Therefore, the plaster must be applied to maximize its strength and its adherence to the bales and/or any wire reinforcement, but also to create the desired finish.

Many guides to the preparation and application of plaster finishes exist, and should be consulted before final selection of materials and application technique. The use of finely chopped straw or poly-fibres in the plaster is highly recommended.

Plaster Reinforcement

Wire mesh has been used widely in straw bale building to provide reinforcement for the plaster skins. However, the nature of the strong bond between the straw bale and the plaster, which accounts for the impressive structural capabilities of bale walls, suggests that the kind of reinforcement offered by steel in other plaster applications is not required for straw bale walls, especially where chopped straw or poly-fibres are added to the plaster to add strength. Despite the numerous buildings created and tests performed without the use of steel reinforcement in the plaster, using a wire reinforcement is still common practice.

The Four Basic Plasters

The four basic plasters are *cement / lime, clay* or *earth-based, lime,* and *gypsum.* The most common plasters are cement-based, creating durable surfaces with impressive strength. Earth-based plasters are usually made of indigenous soil which allow buildings to blend with the local environment, but require frequent resurfacing. Lime plasters have largely been replaced with cement plasters, but they have a long and proven history of working well. Gypsum plaster is the kind often found in older houses, and is the main ingredient in modern drywall.

The Plastering Process

It is standard practice to apply three coats of plaster to both the interior and exterior walls. Together, the three coats should average 1" thick. The first coat is called the *scratch coat,* and is the most time consuming and material intensive coat to apply. It uses a higher amount of binder to aggregate, except in the case of lime plasters. It is called the scratch coat because the surface is literally scratched to form a mechanical grip for the next coat. The next coat is called the *brown coat,* and it will define the overall shape of the wall. It can be lightly scratched with a broom to help bond the final coat. The final coat is is the *colour or finish coat.* It will define the look of the wall. Various textures and colours can be applied from pebbles to light colour washes.

1.0 FOUNDATIONS AND FLOORS

Foundation Systems

Straw bale structures can be built on any commonly used foundation system including poured concrete or concrete block stem walls with or without basements. More often foundations are designed to minimise concrete use and excavation depth. One of the preferred foundation systems is the frost-protected shallow foundation, which requires a depth of only 18" and is protected from potential frost action by extending the insulation out 6" for every foot of frost depth. This system can include an integral floor slab or a separate perimeter beam and slab on grade. Thermal bridging should be given special attention when designing foundations so that the super insulated wall isn't compromised. In-floor hydronic heating systems are recommended as a compliment to straw bale walls.

Structural Requirements

The most important structural requirement in the design of foundations is that the plaster skins bear directly on or transfer loads to the foundation, because it is the plaster skins which transfer the roof loads to the foundation. Wooden floors require blocking under the interior skin.

Moisture Separation

Straw bales must not be in direct contact with concrete foundations. Bales should be elevated a minimum of 1½" above foundations using a curb rail, which in turn should be separated from the concrete surface using foam gaskets and separated from the straw bales using a 6 mil. polyethylene sheet. Suitable flashing must be detailed to prevent wicking of moisture between the bale wall and the foundation via the plaster skin

Experimental Foundations

Many experimental foundations have been successfully implemented including sand bags and rammed earth tires. These foundation systems are not detailed herein. However, one slightly more experimental detail has been included, and that is the rubble trench foundation, which in one detail uses a concrete cap, and in another uses no concrete whatsoever, but rather a slate cap which acts as the moisture break between the foundation and bale wall.

Floors

Floors for straw bale walls include standard wood framed floors, and concrete slab on grade floors, as well as one detail for a brick floor which again uses no concrete at all, and therefore has been drawn together with the rubble trench foundation, but can be used with any foundation type.

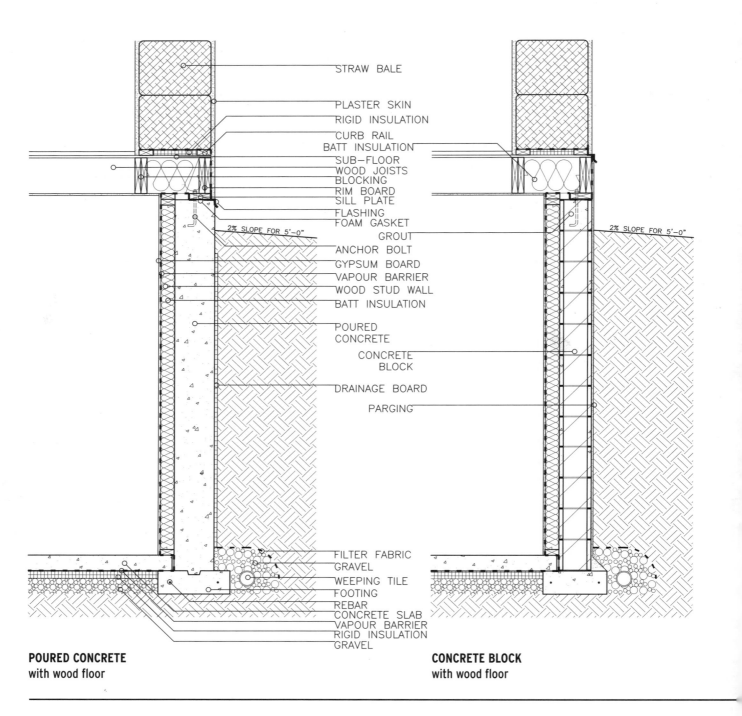

STRAW BALE
PLASTER SKIN
RIGID INSULATION
CURB RAIL
BATT INSULATION
SUB-FLOOR
WOOD JOISTS
BLOCKING
RIM BOARD
SILL PLATE
FLASHING
FOAM GASKET
2% SLOPE FOR 5'-0"
GROUT
ANCHOR BOLT
GYPSUM BOARD
VAPOUR BARRIER
WOOD STUD WALL
BATT INSULATION
POURED CONCRETE
CONCRETE BLOCK
DRAINAGE BOARD
PARGING
FILTER FABRIC
GRAVEL
WEEPING TILE
FOOTING
REBAR
CONCRETE SLAB
VAPOUR BARRIER
RIGID INSULATION
GRAVEL

POURED CONCRETE
with wood floor

CONCRETE BLOCK
with wood floor

NOTES

$1/2$" = 1'- 0"

1. Foundations to be damp-proofed ($3/4$" plastic drainboard typ., or parging)
2. Poured concrete and concrete block walls typically 8" or 10" thick
3. Weeping tile (4" typ.) and reinforcing steel ($1/2$" typ.) per local building codes
4. Depth of wood floor joists dependent on span (11" typ.)
5. Rim board and blocking under plaster skins

1.2 STEM FOUNDATIONS without a basement

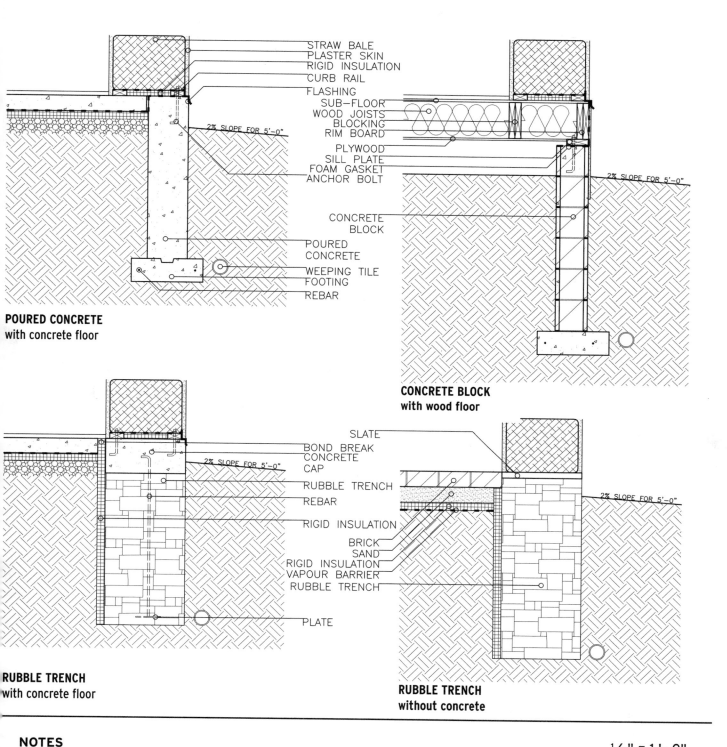

STRAW BALE
PLASTER SKIN
RIGID INSULATION
CURB RAIL
FLASHING
SUB-FLOOR
WOOD JOISTS
BLOCKING
RIM BOARD
PLYWOOD
SILL PLATE
FOAM GASKET
ANCHOR BOLT

2% SLOPE FOR 5'-0"

CONCRETE
BLOCK

POURED
CONCRETE

WEEPING TILE
FOOTING
REBAR

POURED CONCRETE
with concrete floor

CONCRETE BLOCK
with wood floor

SLATE
BOND BREAK
CONCRETE
CAP

2% SLOPE FOR 5'-0"

RUBBLE TRENCH

REBAR

RIGID INSULATION

BRICK
SAND
RIGID INSULATION
VAPOUR BARRIER
RUBBLE TRENCH

PLATE

RUBBLE TRENCH
with concrete floor

RUBBLE TRENCH
without concrete

NOTES

1. Rubble trench as wide as bale wall and plaster (16" or 21" typ.)
2. Depth of wood floor joists dependent on span (11" typ.)
3. Concrete floor typically 4" concrete on 2" insulation on 4" gravel on compacted soil
4. Brick floor typically 4" brick on 4" sand on 2" insulation on compacted soil
5. Welded wire fabric embedded in slab optional (WWF10 6x6 typ.)
6. Crawlspace to be vented to outdoors (2 openings on 2 opposing walls typ.)
7. Brick or tile over sand floor can be used with poured concrete stem wall.

$1/2$" = 1'- 0"

1.3 PIER FOUNDATIONS

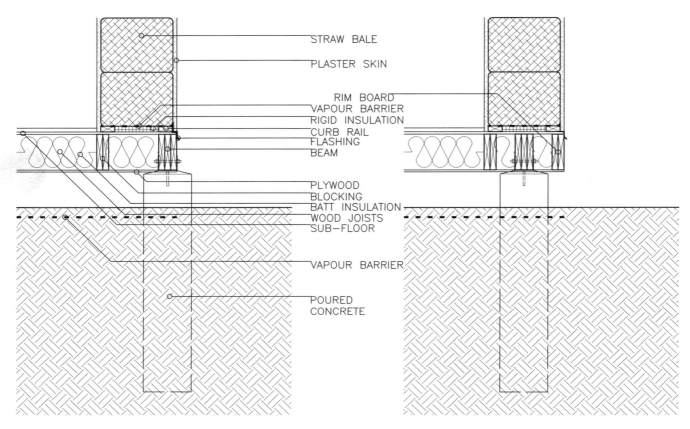

STRAW BALE

PLASTER SKIN

RIM BOARD
VAPOUR BARRIER
RIGID INSULATION
CURB RAIL
FLASHING
BEAM

PLYWOOD
BLOCKING
BATT INSULATION
WOOD JOISTS
SUB-FLOOR

VAPOUR BARRIER

POURED
CONCRETE

PERIMETER BEAM

BEAM CENTRED ON BALE

NOTES

$^1/_2$" = 1'- 0"

1. Piers of poured concrete, concrete block, or treated wood

2. Beam centred on wall or at perimeter

3. Rim board and blocking under plaster skins

4. Weeping tile not required

5. Reinforcing steel dependent on loads

6. Depth of wood floor joists dependent on span (11" typ.)

7. Vapour barrier below grade under building optional

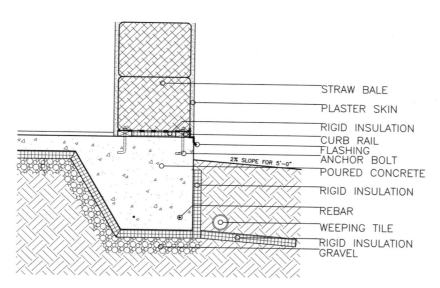

STRAW BALE
PLASTER SKIN
RIGID INSULATION
CURB RAIL
FLASHING
ANCHOR BOLT
POURED CONCRETE
RIGID INSULATION
REBAR
WEEPING TILE
RIGID INSULATION
GRAVEL

2% SLOPE FOR 5'-0"

INTEGRAL SLAB AND BEAM

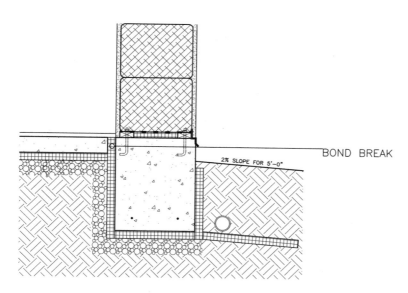

BOND BREAK

2% SLOPE FOR 5'-0"

ISOLATED PERIMETER BEAM

NOTES

1. Weeping tile (4" typ.) and reinforcing steel (1/2" typ.) per local building codes
2. Concrete floor typically 4" concrete on 2" insulation on 4" gravel on compacted soil
3. Welded wire fabric embedded in slab optional (WWF10 6x6 typ.)
4. Rigid insulation to extend horizontally 6" for every foot of frost depth
5. Perimeter beam may be integral with floor slab or isolated by a bond break
6. Anchor bolts may be replaced by concrete nails or screws

1/2" = 1'- 0"

1.5 WOOD FRAMED FLOORS

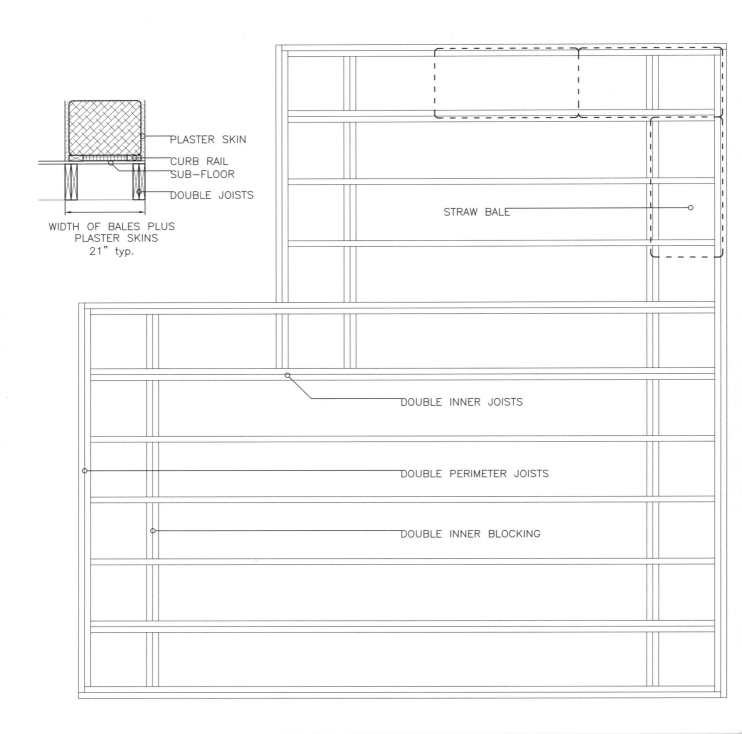

PLASTER SKIN

CURB RAIL

SUB-FLOOR

DOUBLE JOISTS

WIDTH OF BALES PLUS
PLASTER SKINS
21" typ.

STRAW BALE

DOUBLE INNER JOISTS

DOUBLE PERIMETER JOISTS

DOUBLE INNER BLOCKING

NOTES $1/2" = 1'-0"$

1. Double inner joists
2. Joist depths dependent on span (11" typ.)
3. Joist spacing dependent on loads (16" typ.)
4. Bale width — 14" or 19" typ., plaster width — 1" typ. per side. Total wall width — 16" or 21" typ.

1.6 CONCRETE FLOORS

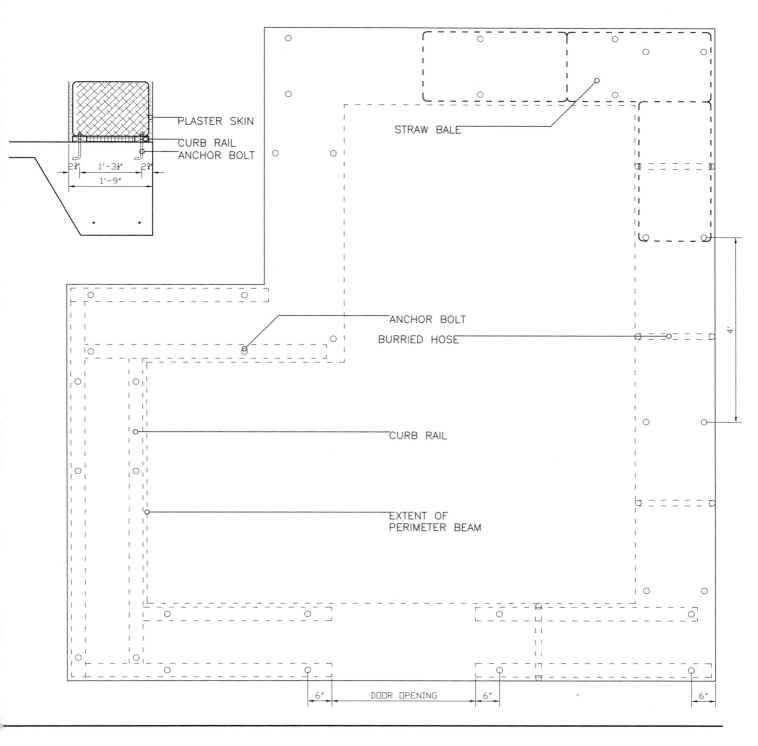

PLASTER SKIN

CURB RAIL
ANCHOR BOLT

2¾" 1'-3½" 2¾"
1'-9"

STRAW BALE

ANCHOR BOLT
BURRIED HOSE

4'

CURB RAIL

EXTENT OF
PERIMETER BEAM

6'   DOOR OPENING   6'   6'

½" = 1'- 0"

NOTES

1. Anchor bolts spaced at 4' o.c.; concrete nails spaced at 2' o.c.
2. Anchor bolts to be placed within 2' of either side of door openings
3. Anchor bolts to be placed within 2' of corners
4. Pre-stressing bolts, rebar, or hose should be set in concrete floor
5. Pre-stressing bolts, rebar, or hose spaced at 4' o.c., minimum 2 per wall
6. Pre-stressing bolts, rebar, or hose to be placed within 2' of corners
7. Curb rails staggered at corners

2.0 CURB RAILS

Curb Basics

In order to elevate the bale wall above potential pooling of water on the floor, and to provide a nailing surface for stucco mesh and/or finishing trim, bale curbs are typically used between the foundation and the first course of bales. Spacing of foundation curbs must correspond to actual bale dimensions which are typically 19" on flat and 14" on edge for two-string bales. Curb rails are typically of 2x4 construction. Curb rails should be set 1" in from the exterior face of foundation walls and 1" from the interior face of perimeter beams to accommodate the 1" required for the plaster skins, or else the curbs should be of 2x6 construction and set flush to the face of foundation walls to allow the plaster skins to bear on the curbs.

Curb Details

Curbs may be made of non-structural grade lumber or plastic wood. The void between rails is to be filled with rigid insulation including rigid foam, rock wool, perlite, or drainboard. A 6 mil. polyethylene vapour barrier is required between the curb rail and straw bales. Curb rails should also be separated from concrete surfaces by foam gaskets. To further protect the wicking of moisture, tar paper or asphalt emulsion may be placed between concrete foundations and curb rails. Curb rails can be attached to the foundation with anchor bolts, nails, or screws at sizes and intervals required by local codes.

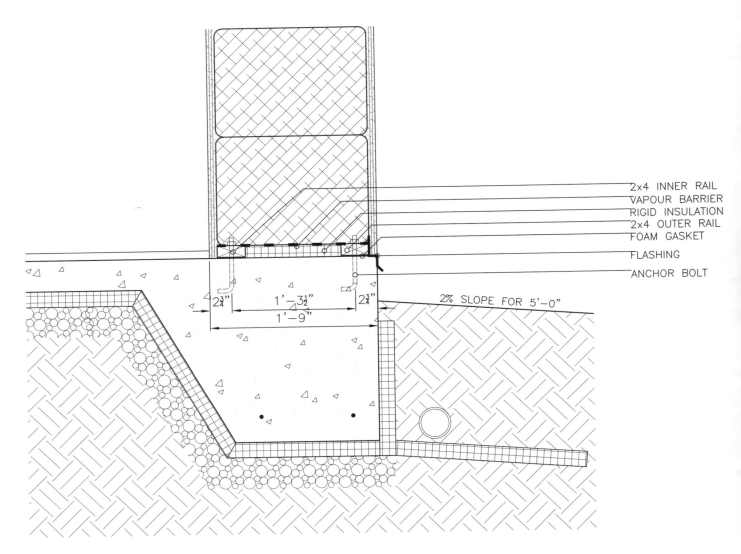

2x4 INNER RAIL
VAPOUR BARRIER
RIGID INSULATION
2x4 OUTER RAIL
FOAM GASKET
FLASHING
ANCHOR BOLT

2% SLOPE FOR 5'–0"

$2\frac{3}{4}$"   1'–$3\frac{1}{2}$" $2\frac{3}{4}$"
1'–9"

PLASTER SKINS BEARING DIRECTLY ON FOUNDATION

1" = 1'– 0"

NOTES

1. 2–2x4 rails set apart 19" outside to outside (14" for bales on edge)
2. Exterior rail set 1" from edge of foundation
3. Non-structural grade wood or plastic wood acceptable
4. Rails placed on top of foam gaskets or equivalent
5. Between rails filled with rigid insulation (1"), drainboard, or equivalent
6. Vapour barrier cut flush to outer edges of rails
7. Asphalt emulsion or tar paper between concrete and curb rail optional
8. 3" concrete nail can substitute anchor bolts

2.2 CURB RAILS with a concrete floor

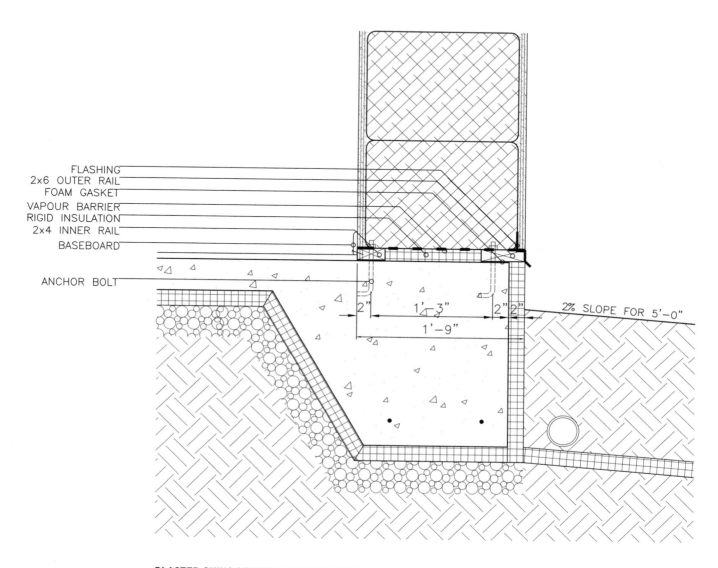

FLASHING
2x6 OUTER RAIL
FOAM GASKET
VAPOUR BARRIER
RIGID INSULATION
2x4 INNER RAIL
BASEBOARD

ANCHOR BOLT

2"   1'-3"   2" 2"

1'-9"

2% SLOPE FOR 5'-0"

PLASTER SKINS BEARING ON CURB RAILS

NOTES

1" = 1'- 0"

1. 2x4 interior rail and 2x6 exterior rail set apart 21" outside to outside (16" for bales on edge)
2. Exterior rail overhangs edge of foundation by thickness of perimeter insulation (2" typ.)
3. Both rails set 1" wider than bale to support plaster skin
4. Baseboards nailed to inner rail
5. Exposed exterior rigid insulation can be parged or covered with hard board
6. 3" concrete nail can substitute anchor bolts
7. Asphalt emulsion or tar paper between concrete and curb rail optional

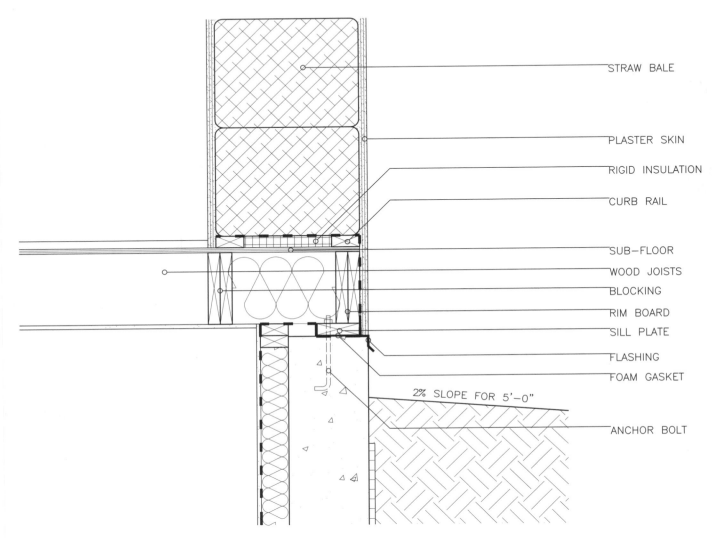

STRAW BALE

PLASTER SKIN

RIGID INSULATION

CURB RAIL

SUB-FLOOR

WOOD JOISTS

BLOCKING

RIM BOARD

SILL PLATE

FLASHING

FOAM GASKET

2% SLOPE FOR 5'-0"

ANCHOR BOLT

PLASTER SKIN BEARING DIRECTLY ON FOUNDATION

1" = 1'- 0"

NOTES

1. 2-2x4 rails set apart 19" outside to outside (14" for bales on edge)
2. Exterior rail set flush with edge of rim board
3. Sill plate and rim board placed 1" from edge of foundation
3. Non-structural grade wood or plastic wood acceptable for curb rails
4. Rails placed on top of foam gaskets or equivalent
5. Between rails filled with rigid insulation (1"), drainboard, or equivalent
6. Vapour barrier cut flush to edge of interior rail

2.4 CURB RAILS with a wood floor

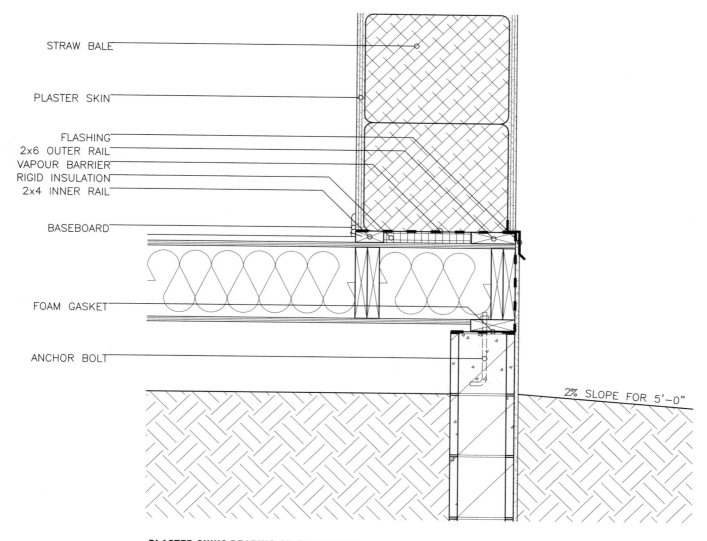

STRAW BALE

PLASTER SKIN

FLASHING
2x6 OUTER RAIL
VAPOUR BARRIER
RIGID INSULATION
2x4 INNER RAIL

BASEBOARD

FOAM GASKET

ANCHOR BOLT

2% SLOPE FOR 5'-0"

PLASTER SKINS BEARING ON CURB RAILS

NOTES

1" = 1'- 0"

1. 2x4 interior rail and 2x6 exterior rail set apart 21" outside to outside (16" for bales on edge)
2. Exterior rail flush with edge of rim board
3. Both rails set 1" wider than bale to support plaster skin
4. Economy grade wood or plastic wood acceptable for curb rails
5. Rails placed on top of foam gaskets or equivalent
6. Between rails filled with rigid insulation ($1\frac{1}{2}$"), drainboard, or equivalent
7. Vapour barrier cut flush to edge of interior rail

3.0 WALLS

Wall Systems

The two types of straw bale walls are load-bearing, often referred to as *Nebraska Style*, and infill walls. Load-bearing walls best utilise the bales as both super insulative and structural. However, height and load limitations do exist. For two-storey buildings, a post and beam structure is often used with the bales as infill elements. Post and beam structures can either be of timber or box beam construction.

Load-Bearing Walls

Load-bearing walls utilise the full strength of the plastered bales in place of structural framing. For this reason, they represent significant savings in materials and labour. Designers must familiarise themselves with the principles and capabilities of load-bearing walls, including the pre-stressing of walls, the placement of openings and allowable loads.

Infill Walls

Straw bales can be used as an insulative infill with a variety of different framing systems, from timber framing to prefabricated steel. Careful attention must be give to the interface between the bales and the framing system to avoid potential problems with air and water leakage. Designers of infill walls can take advantage of the inherent compressive and racking strength of plastered straw bale walls when designing the accompanying structure.

Box Beams

Box beams in this context refer to columns, top plates, and rafters. A box beam is a built-up element of a 2x4 frame sandwiched between plywood sheets and filled with batt insulation. The advantage of box beams is that they are light weight and thin, so they do not interfere with the bale wall or create difficult voids. Box beams can also be made for two storey construction with notches to support the second storey floor framing system.

Moisture Retention

In post and beam timber construction, wherever a post or beam can act as a transfer of water vapour from the indoor environment to the core of the bale wall, a vapour barrier should be used. This scenario occurs when the posts and beams are flush to the interior surface of the wall or slightly inset within the wall. If the posts or beams are wholly within the wall or completely independent of the wall, a vapour barrier is not required. Posts independent of the wall should be separated by at least 6" to allow space for plastering the wall. Framing members should not protrude from the plaster on the exterior of the building, unless well protected by a porch or large roof overhang.

Air Leakage

Wherever plaster meets a framing member, provision must be made to avoid eventual air leakage through the seam created at the intersection. Vapour barrier should be used behind posts and beams, and should extend a minimum of 2" over the straw and behind the plaster and reinforcement wire. Seams between plaster and framing members should be caulked after the plaster is fully cured.

3.1 TEMPORARY BRACING

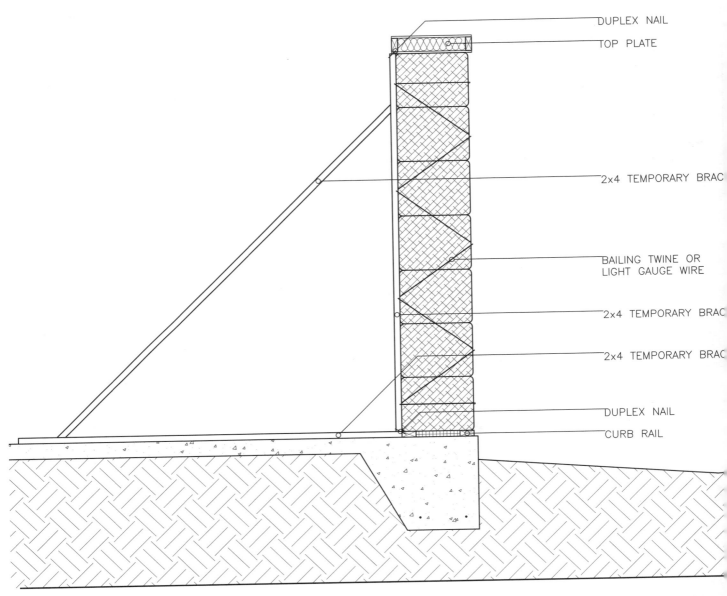

DUPLEX NAIL

TOP PLATE

2x4 TEMPORARY BRAC

BAILING TWINE OR
LIGHT GAUGE WIRE

2x4 TEMPORARY BRAC

2x4 TEMPORARY BRAC

DUPLEX NAIL

CURB RAIL

1/2" = 1'- 0"

NOTES

1. Bales threaded together with a baling needle
2. Braces of 2x4 construction
3. Braces spaced two per wall or approximately every 15'
4. Braces nailed to interior curb rail
5. Braces may be erected prior to wall construction to act as levelling guides
6. Vertical brace member nailed to top plate

3.2 EXTERNAL PINNING (optional)

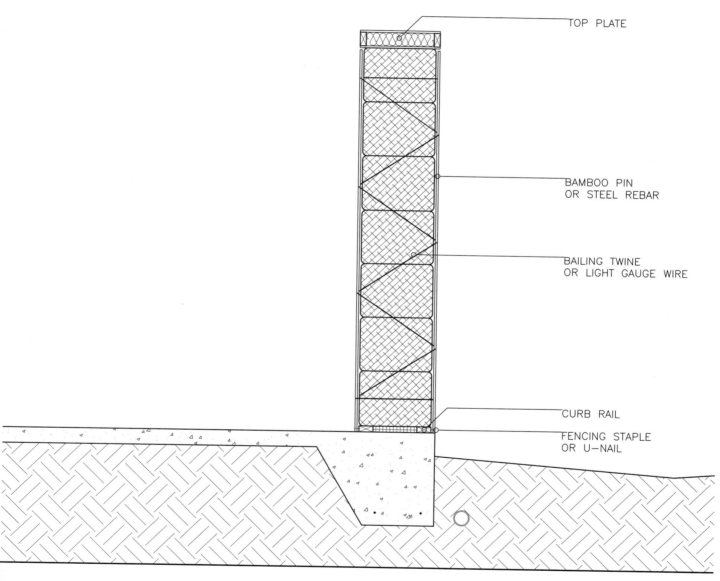

TOP PLATE

BAMBOO PIN
OR STEEL REBAR

BAILING TWINE
OR LIGHT GAUGE WIRE

CURB RAIL

FENCING STAPLE
OR U—NAIL

$1/2" = 1'- 0"$

NOTES

1. Pins of galvanised concrete block ties or bamboo ($1/2"$ typ.)
2. Pins threaded tightly to either side of wall
3. Twine or wire threaded with a baling needle
4. Stop pins 2" from top of top plate
5. Pins remain and get embedded in plaster skins

3.3 POST AND BEAM

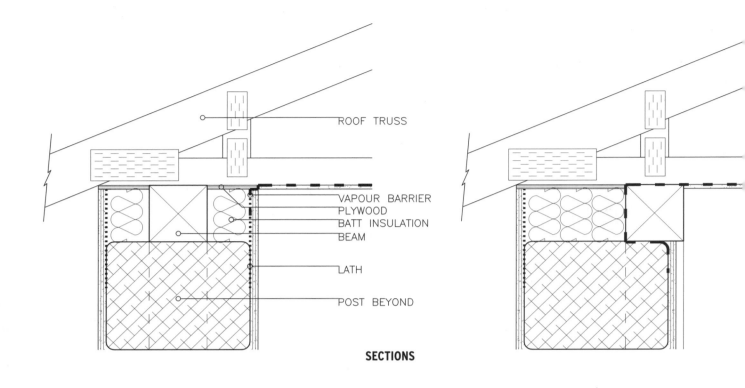

ROOF TRUSS

VAPOUR BARRIER
PLYWOOD
BATT INSULATION
BEAM

LATH

POST BEYOND

SECTIONS

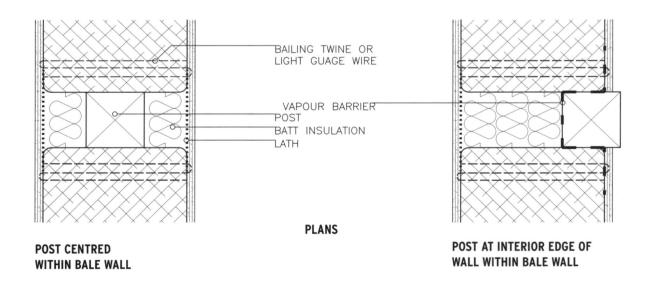

BAILING TWINE OR
LIGHT GUAGE WIRE

VAPOUR BARRIER
POST
BATT INSULATION
LATH

PLANS

**POST CENTRED
WITHIN BALE WALL**

**POST AT INTERIOR EDGE OF
WALL WITHIN BALE WALL**

NOTES

1" = 1'- 0"

1. Batt insulation or packed straw used to fill behind posts
2. Lath stitched on with wire or twine and a baling needle
3. Lath to overlap bales a minimum of 2"
4. Vapour barrier required behind posts and beams when situated at interior edge of wall
5. Vapour barrier tied into ceiling vapour barrier

3.4 POST AND BEAM

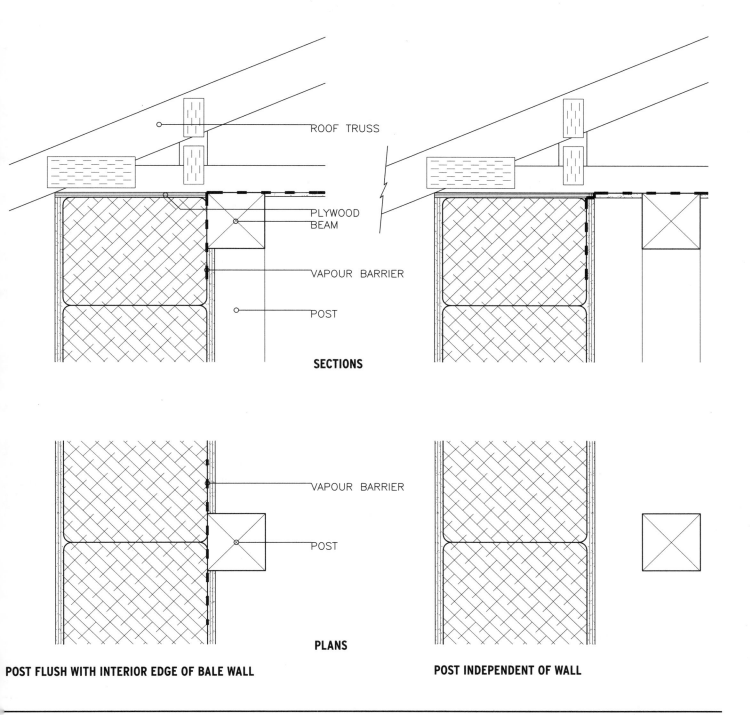

ROOF TRUSS

PLYWOOD
BEAM

VAPOUR BARRIER

POST

SECTIONS

VAPOUR BARRIER

POST

PLANS

POST FLUSH WITH INTERIOR EDGE OF BALE WALL

POST INDEPENDENT OF WALL

NOTES   1" = 1 '- 0"

1. Lath stitched on with wire or twine and a baling needle
2. Lath to overlap bales a minimum of 2"
3. Vapour barrier required behind posts and beams when situated flush with interior edge of bale wall
4. Posts independent of wall separated from wall a minimum of 6" to facilitate proper plastering
5. Independent posts on the exterior require large roof overhangs

3.5 BOX BEAM CONSTRUCTION

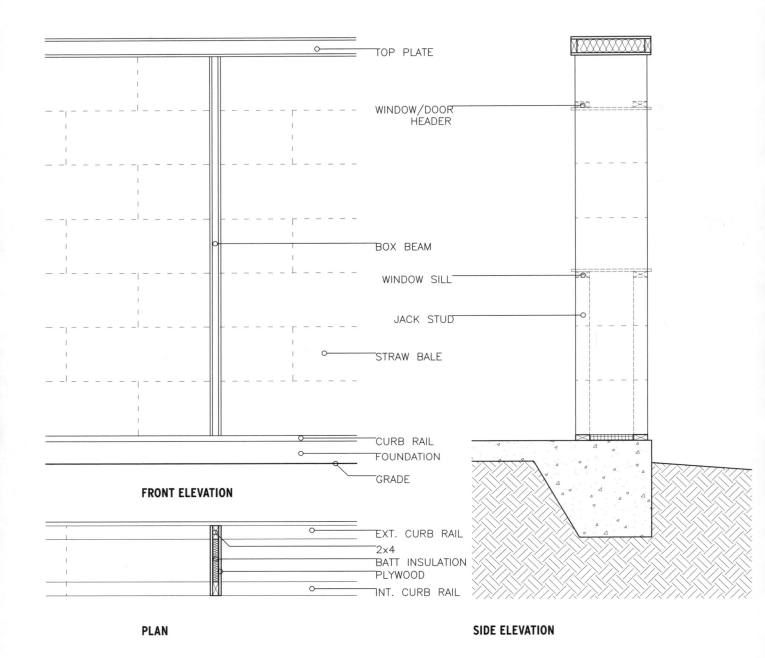

TOP PLATE

WINDOW/DOOR
HEADER

BOX BEAM

WINDOW SILL

JACK STUD

STRAW BALE

CURB RAIL
FOUNDATION
GRADE

FRONT ELEVATION

EXT. CURB RAIL
2x4
BATT INSULATION
PLYWOOD
INT. CURB RAIL

PLAN

SIDE ELEVATION

NOTES $1/2" = 1'- 0"$

1. Box beams of 2x4 and plywood ($1/2"$ typ.) construction
2. Fill box beams with batt insulation before nailing closed
3. For 2 storey construction, box beams can be notched to accept floor beam
4. Jack studs (2x4s) required under window sills

3.6 BOX BEAM CONSTRUCTION

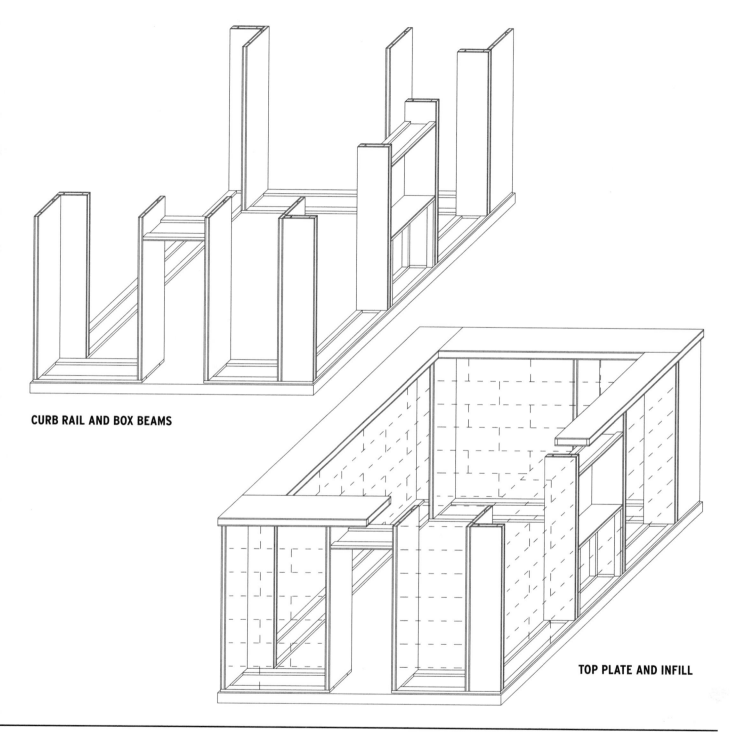

CURB RAIL AND BOX BEAMS

TOP PLATE AND INFILL

NOTES

NOT TO SCALE

1. 2x4 blocking between curb rails under box beams
2. Corners of buildings constructed of two box beams at right angles

4.0 OPENINGS

Rough Framing

Rough framing will determine the final plaster finish. Consideration for eventual 1" plaster thickness should be made. Lintel requirements are not identical to conventional framing. The structure of the top plate and the depth of the plaster lintel over openings reduces the need for window or door bucks to carry full loads. 2x4 or 2x6 lintels on inner and outer edge of rough frames is typical for openings wider than 48". Where plaster is to be applied to the window or door buck, diamond lath over tar paper is required.

Moisture Protection

Windows must incorporate sills with a drip kerf or equivalent flashing. Drip edge or flashing must protrude a minimum of ½" beyond the finished plaster surface. Windows incorporating a full exterior flange are suitable. Drip edge flashing must be incorporated over windows and doors if unprotected by roof overhangs. Windows must not have the exterior sill of unprotected plastered bales. Therefore, windows should be placed close to or flush with the outside face of the wall.

Window and Door Bucks

Window and door bucks are typically of 2x4 and plywood construction. For openings greater than 48", the 2x4s should be placed on edge. 2x4 supports can extend to the curb to support buck, or buck can float in bales. 2x4 jack studs should be used to support the rough sill, unless the buck is designed to float in the wall. The plywood attached to the inside of the 2x4 framing should extend 1" wider than the frame on all sides to act as a plaster stop. If wooden trim is to be applied, use a 1x2 nailer or set the buck flush to the surface of wall. For curved plaster finishes, form curves using diamond lath.

Cutting Window After Plastering

Recent experiments in cutting out window openings after plastering have shown that savings in time and lumber can be achieved. Only door openings and large windows are required to be framed in before wall construction and plastering. After applying the *scratch coat*, a kerf can be cut into the interior and exterior plaster skins using a diamond blade in a circular saw. Afterwards, a hay saw or chainsaw can be used to cut through the straw. The block of plastered straw can then be removed and a light-weight 1" lumber frame can be inserted into the opening and dowelled into place. The window can then be attached to this frame. Openings up to 8 feet wide have been cut into walls in this manner with no visible compromise evident in the wall. The *finish coat* of plaster is used to create the desired finish around the opening.

4.1 OPENINGS

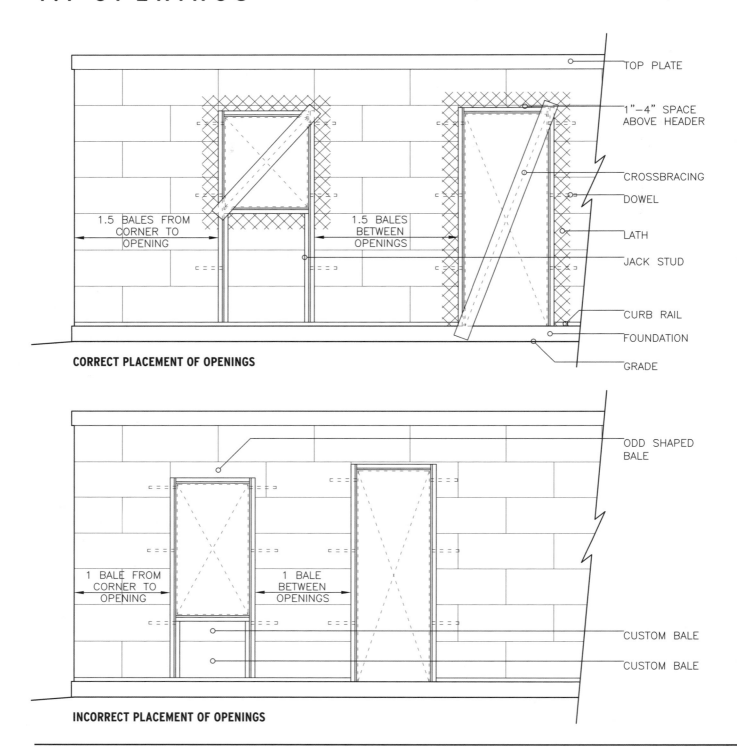

CORRECT PLACEMENT OF OPENINGS

TOP PLATE

1"–4" SPACE ABOVE HEADER

CROSSBRACING

DOWEL

LATH

JACK STUD

CURB RAIL

FOUNDATION

GRADE

1.5 BALES FROM CORNER TO OPENING

1.5 BALES BETWEEN OPENINGS

INCORRECT PLACEMENT OF OPENINGS

ODD SHAPED BALE

1 BALE FROM CORNER TO OPENING

1 BALE BETWEEN OPENINGS

CUSTOM BALE

CUSTOM BALE

NOTES

NOT TO SCALE

1. Sides of bucks extend to curb rails
2. Custom and odd-shaped bales should be avoided
3. Min. 1 bales from an outside corner to an opening
4. Min. 1 bales between openings
5. Wrap openings with lath extending 6" over bale, all around
6. Lath stitched with bailing twine or light gauge wire
7. Dowels optional

4.2 WINDOW BUCKS

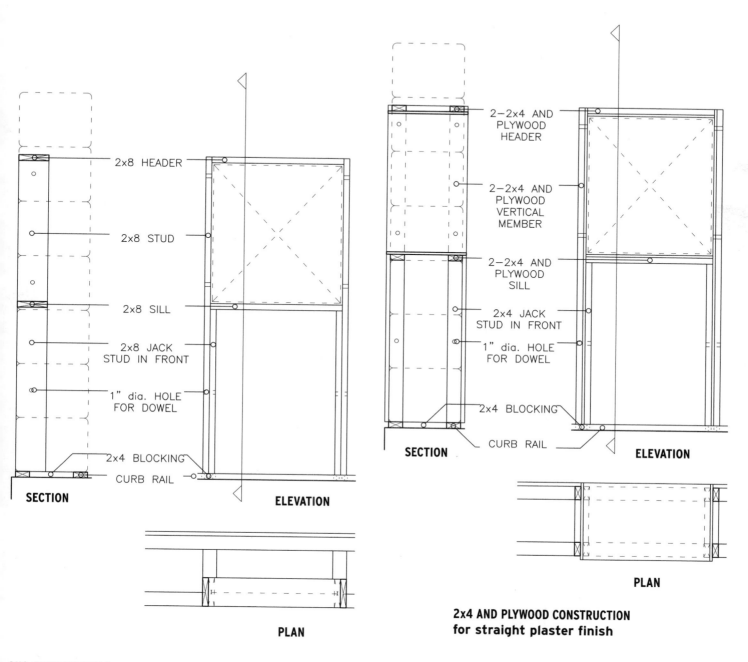

SECTION — 2x8 HEADER, 2x8 STUD, 2x8 SILL, 2x8 JACK STUD IN FRONT, 1" dia. HOLE FOR DOWEL, 2x4 BLOCKING, CURB RAIL

ELEVATION

PLAN

2X8 CONSTRUCTION
for curved plaster finish

2-2x4 AND PLYWOOD HEADER, 2-2x4 AND PLYWOOD VERTICAL MEMBER, 2-2x4 AND PLYWOOD SILL, 2x4 JACK STUD IN FRONT, 1" dia. HOLE FOR DOWEL, 2x4 BLOCKING, CURB RAIL

SECTION

ELEVATION

PLAN

2x4 AND PLYWOOD CONSTRUCTION
for straight plaster finish

NOTES

1. Bucks for curved plaster finish of 2x8 construction
2. Bucks for straight plaster finish of 2x4 and plywood construction
3. Jack studs required under window sills
4. Plywood set 1" beyond either stud to act as plaster stop
5. 2x4 blocking required between curb rails under studs
6. Dowels optional

$1/2" = 1'- 0"$

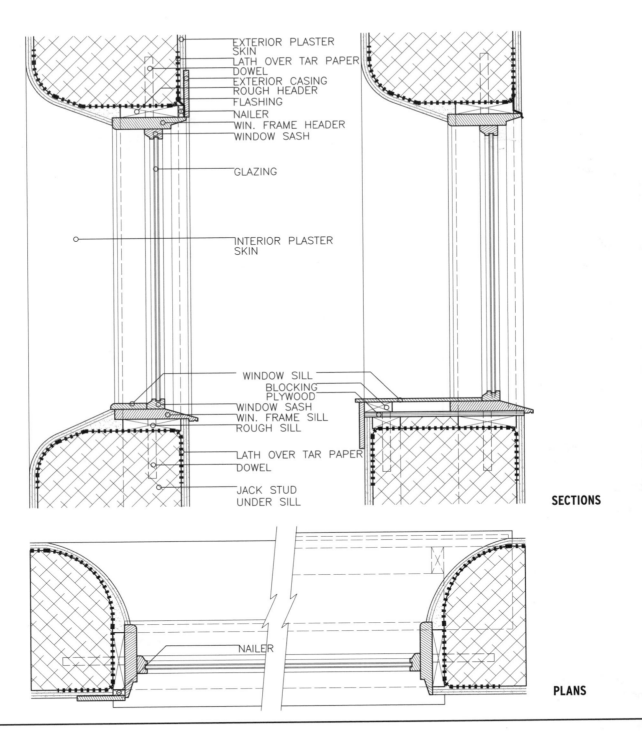

EXTERIOR PLASTER SKIN
LATH OVER TAR PAPER
DOWEL
EXTERIOR CASING
ROUGH HEADER
FLASHING
NAILER
WIN. FRAME HEADER
WINDOW SASH

GLAZING

INTERIOR PLASTER SKIN

WINDOW SILL
BLOCKING
PLYWOOD
WINDOW SASH
WIN. FRAME SILL
ROUGH SILL

LATH OVER TAR PAPER

DOWEL

JACK STUD UNDER SILL

SECTIONS

NAILER

PLANS

1" = 1'- 0"

NOTES

1. 1x2 nailers required for attaching trim or extend studs 1" beyond bail wall

2. Window frame sill at exterior face of wall

3. Sill should extend minimum $1/2$" beyond exterior plaster skin

4. Interior plaster applied directly up to window sill

5. Lath used to make curves

6. Lath over tar paper for moisture control

4.4 WINDOW DETAILS for straight plaster finish

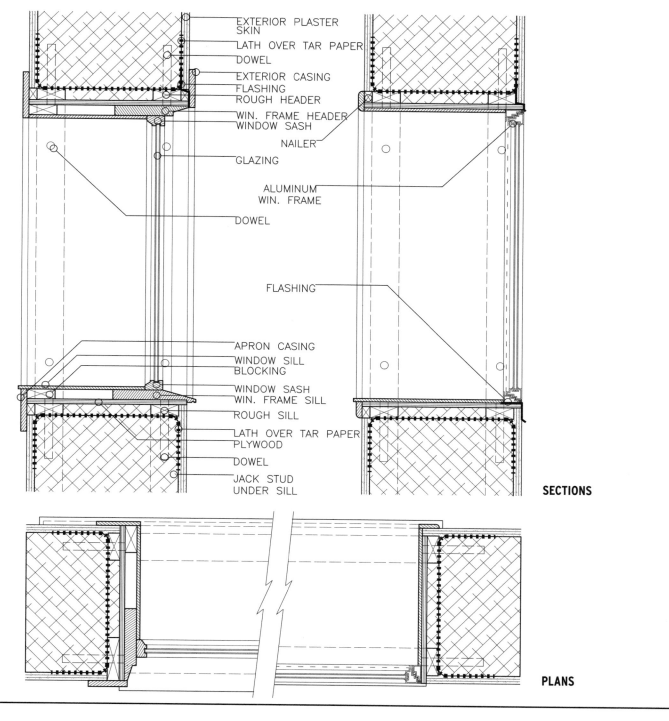

EXTERIOR PLASTER SKIN
LATH OVER TAR PAPER
DOWEL
EXTERIOR CASING
FLASHING
ROUGH HEADER
WIN. FRAME HEADER
WINDOW SASH
NAILER
GLAZING
ALUMINUM WIN. FRAME
DOWEL
FLASHING
APRON CASING
WINDOW SILL
BLOCKING
WINDOW SASH
WIN. FRAME SILL
ROUGH SILL
LATH OVER TAR PAPER
PLYWOOD
DOWEL
JACK STUD UNDER SILL

SECTIONS

PLANS

NOTES

1. 1x2 nailers required for attaching trim or extend studs 1" beyond bail wall
2. Aluminum frames flush or 1/2" beyond exterior plaster skin
3. Plywood extends 1" beyond surface of bales to act as a plaster stop
4. Lath over tar paper for moisture control

1" = 1'- 0"

5.0 TOP PLATES

Top plates must provide for transfer of roof loads to the plaster skins, and be able to carry loads over wide openings in bale walls. Therefore, they should project 1" beyond the face of the bales. They must also prevent thermal bridging and are therefore typically filled with batt insulation. Rock wool, cellulose, rigid foam, loose straw or an equivalent may also be used. If blocking is required in the design, it should be spaced at the same centres as the roof framing and where the pre-stressing loads are to be applied.

The four main top plate designs are the *box beam, plywood over beam, modified ladder*, and *concrete top plate*. Box beams should have the parallel 2x4 members staggered. These members should also be doubled up when spanning large openings. Plywood over beam top plates are constructed of ⅝" or ¾" plywood over a beam of solid lumber or laminate centred on the wall and of dimensions defined by local requirements. Diamond lath spans the hollow between the top of the bale wall and the plywood. This hollow is stuffed with batt insulation or equivalent. Modified ladder plates are similar to box beams except the outer rails are typically of 2x6 or 2x8 members which extend down the sides of the wall for added stability. Concrete top plates are poured directly on the top course of bales. The pre-stressing wire is cast in the top plate, and the 2x4 forms remain in place. Insulation is placed between the forms and concrete. 2x4 blocking should be attached at the same centres as the roof framing.

5.1 TOP PLATES

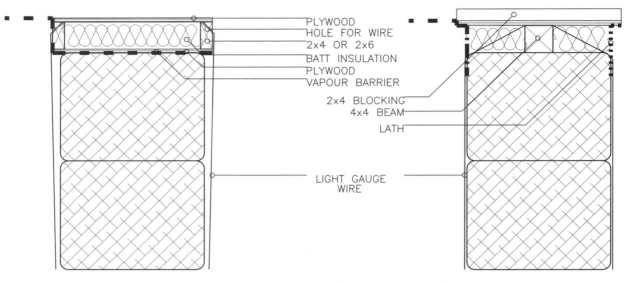

PLYWOOD
HOLE FOR WIRE
2x4 OR 2x6
BATT INSULATION
PLYWOOD
VAPOUR BARRIER

2x4 BLOCKING
4x4 BEAM
LATH

LIGHT GAUGE
WIRE

BOX BEAM TOP PLATE   **PLYWOOD OVER BEAM TOP PLATE**

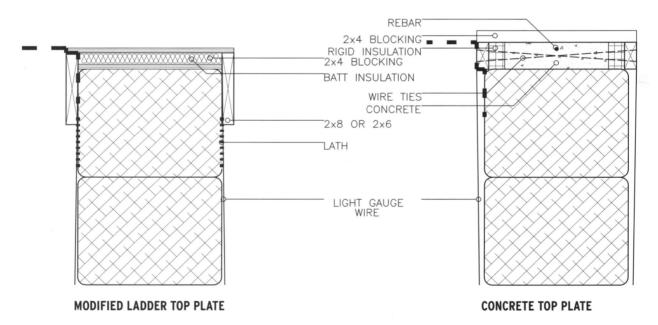

REBAR
2x4 BLOCKING
RIGID INSULATION
2x4 BLOCKING
BATT INSULATION

WIRE TIES
CONCRETE

2x8 OR 2x6

LATH

LIGHT GAUGE
WIRE

MODIFIED LADDER TOP PLATE   **CONCRETE TOP PLATE**

NOTES

1. Top plates extend 1" beyond face of bales
2. Ceiling vapour barrier tied into top plate
3. Top plates insulated with loose straw, cellulose, or mineral wool
4. Lath should extend 6" over straw bales
5. Lath attached with bailing twine or light gauge wire
6. Concrete poured directly on bales
7. Blocking spaced to match roof structure

1" = 1'- 0"

5.2 TOP PLATES

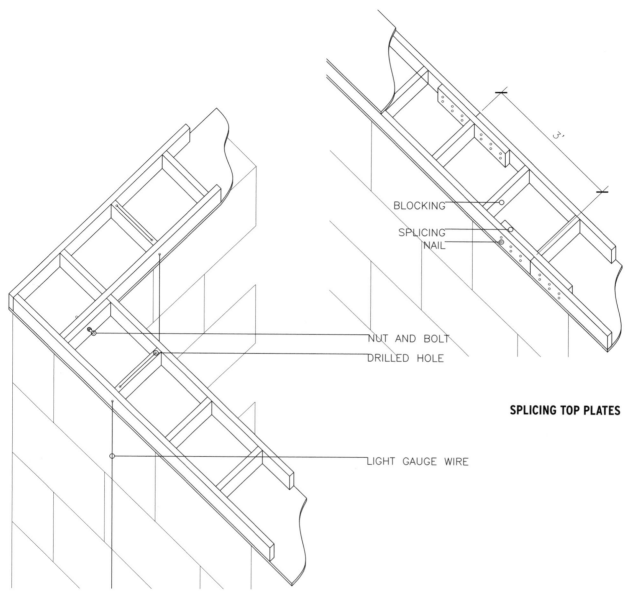

BLOCKING

SPLICING
NAIL

3'

NUT AND BOLT

DRILLED HOLE

SPLICING TOP PLATES

LIGHT GAUGE WIRE

CORNER BRACING

NOTES

1. Joints in top plate rails staggered minimum 36"
2. Joints scabbed with nails and adhesive
3. Corners blocked and bolted
4. Poles drilled in rail and blocking for pre-stressing wire
5. Top plates filled with batt insulation or equivalent

$^1/_2$" = 1 '– 0"

6.0 PRE-STRESSING

Load-bearing Wall System

For load-bearing walls, pre-stressing and levelling is achieved mechanically by using either a 9–12 gauge galvanised wire or a threaded rod. The *wire method* includes either wrapping the wire over the top plate and through the floor, requiring a single connection, or by wrapping one wire over the top plate and another two wires embedded in the foundation at both the exterior and interior, requiring 2 connections. The wire connections are made by looping each end and linking them together. The loops are created with a saddle clamp. The compression is achieved by either a 2 tonne capacity come-along or the more low tech equivalent: two vise grips and a steel T-section. The principal is the same for both. Pressure is applied upwards on the lower wire and downwards on the upper wire until the desired compression is achieved. The *threaded rod method* uses threaded rod at either side of the wall tight to the surface of the wall. The top of the rods penetrate blocking on the top plate. A washer and nut when tightened achieve the required compression. Pre-stressing elements should be located at 4-foot centres and at 1' of either side of openings greater than 4'. Walls are typically compressed 1"–6" to achieve a level top plate. Sharp bends in the wire should be avoided.

Infill Wall System

No pre-stressing of straw bale walls is required when framing is used.

6.1 PRE-STRESSING

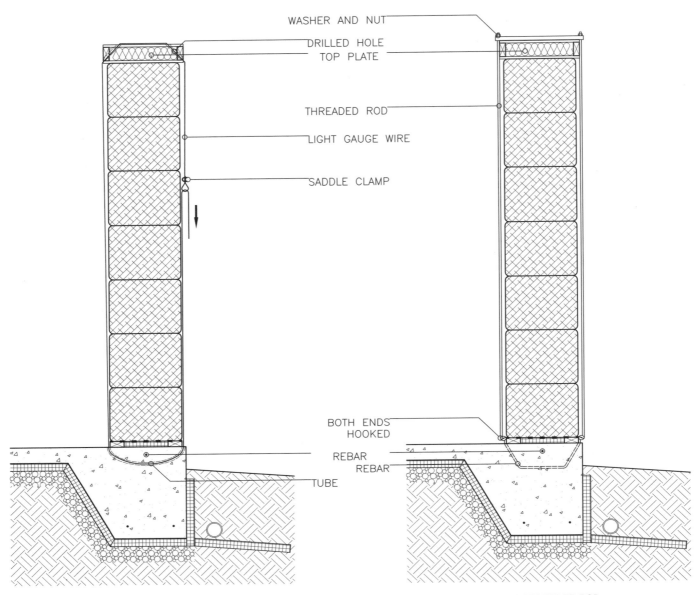

WASHER AND NUT
DRILLED HOLE
TOP PLATE

THREADED ROD

LIGHT GAUGE WIRE

SADDLE CLAMP

BOTH ENDS HOOKED

REBAR
REBAR

TUBE

PRE-STRESSING WITH WIRE

PRE-STRESSING WITH THREADED ROD

$^{1}/_{2}$" = 1'- 0"

NOTES

1. Loop created with a saddle clamp
2. Alternate wire tensioning between inside and outside
3. 9-12 gauge galvanise wire at 4' o.c. and at 2' from corners, windows and doors
4. Force applied in downward direction to compress wall
5. Holes in top plate drilled at 45° for smooth routing of wires
6. Threaded rod tight to surface of bales
7. Washer and nut over blocking to compress wall

6.2 PRE-STRESSING

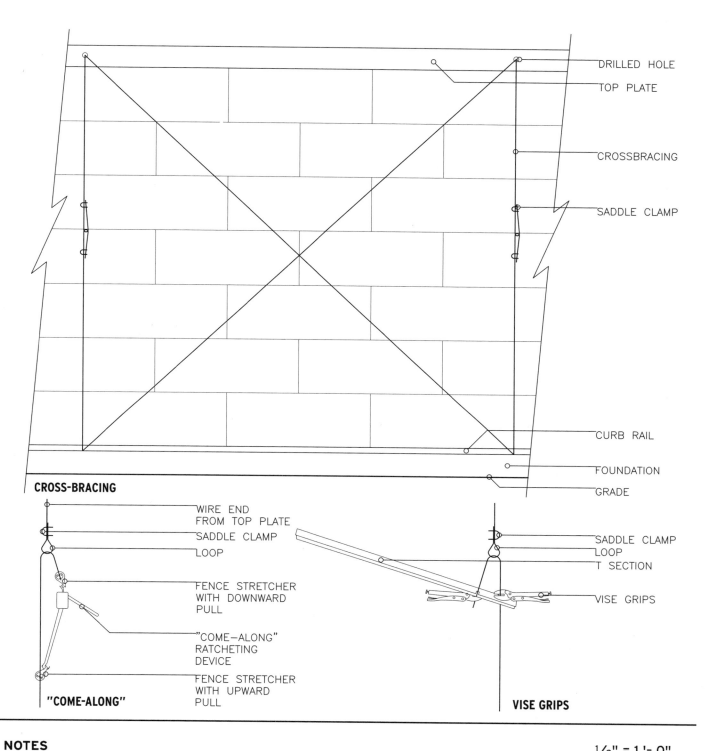

DRILLED HOLE
TOP PLATE
CROSSBRACING
SADDLE CLAMP
CURB RAIL
FOUNDATION
GRADE

CROSS-BRACING

WIRE END
FROM TOP PLATE
SADDLE CLAMP
LOOP

FENCE STRETCHER
WITH DOWNWARD
PULL

"COME—ALONG"
RATCHETING
DEVICE

FENCE STRETCHER
WITH UPWARD
PULL

"COME-ALONG"

SADDLE CLAMP
LOOP
T SECTION

VISE GRIPS

VISE GRIPS

NOTES

1. 45° holes can be drilled to prevent sharp bends
2. 2x4 blocking glued and screwed to joist or blocking to guide wires
3. Loops made with saddle clamps
4. cross-bracing suggested at one location on each wall
5. T-section notched in two locations to accept wires

$1/2" = 1'- 0"$

6.3 PRE-STRESSING

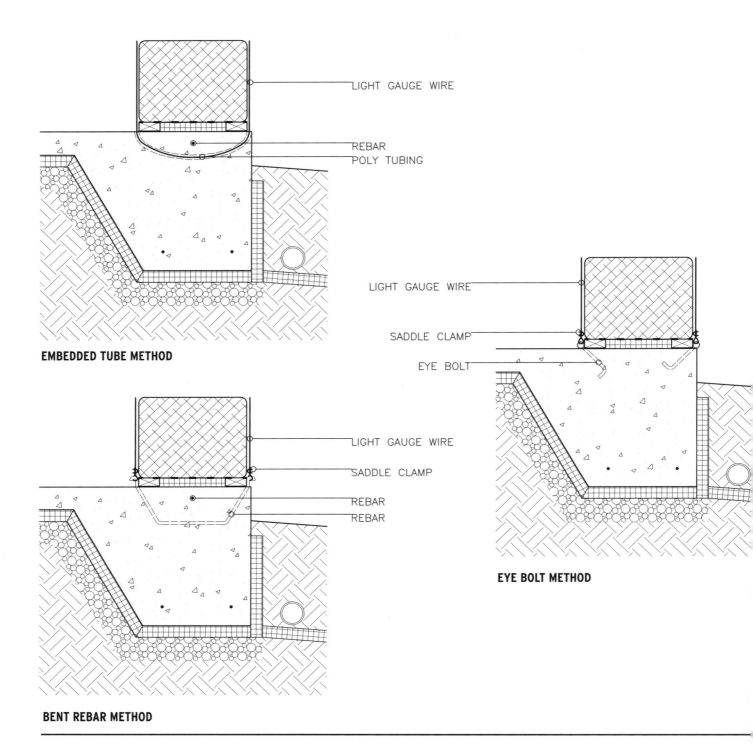

LIGHT GAUGE WIRE

REBAR
POLY TUBING

EMBEDDED TUBE METHOD

LIGHT GAUGE WIRE

SADDLE CLAMP

EYE BOLT

EYE BOLT METHOD

LIGHT GAUGE WIRE

SADDLE CLAMP

REBAR
REBAR

BENT REBAR METHOD

NOTES

1. Tubing, bent rebar or eye bolts embedded in concrete
2. Pre-stressing wire of 9–12 gauge galvanised wire
3. Rebar to span tubing or bent rebar
4. Bent rebar hooked to accept wire loop

$3/4" = 1'- 0"$

6.4 PRE-STRESSING

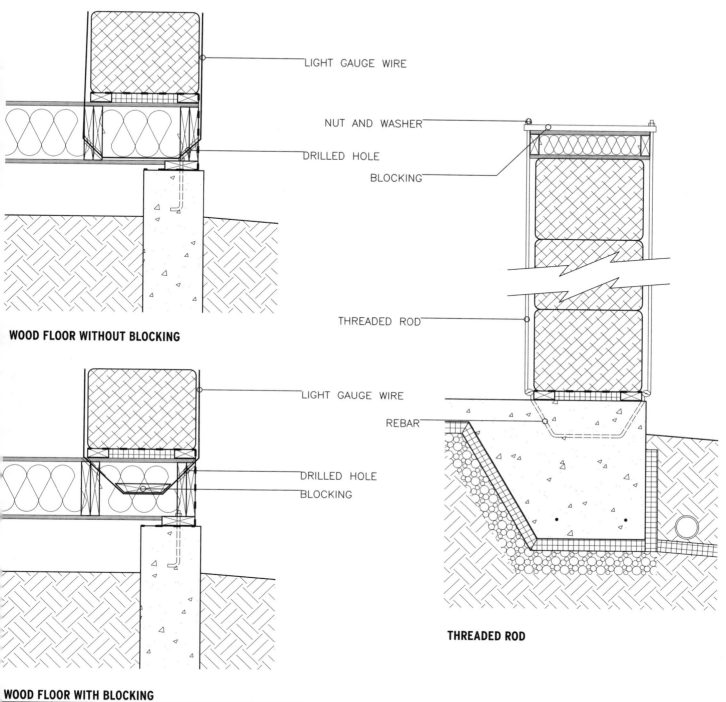

LIGHT GAUGE WIRE

NUT AND WASHER

DRILLED HOLE

BLOCKING

WOOD FLOOR WITHOUT BLOCKING

THREADED ROD

LIGHT GAUGE WIRE

REBAR

DRILLED HOLE

BLOCKING

THREADED ROD

WOOD FLOOR WITH BLOCKING

NOTES

1. 45° holes can be drilled to prevent sharp bends
2. Wire looped around double rim board and double blocking
3. Optional 2x4 blocking glued and screwed to joist to guide wires
4. Threaded rod hooked at bottom
5. Blocking over top plate drilled to accept threaded rod
6. Nut and washer tightened to compress wall

$3/4" = 1'- 0"$

7.0 ROOFS

Roof framing can be used as the protection from exterior wetting recommended by moisture tests on bale walls. Roof overhangs of 18"–24" are adequate protection for single storey walls which are not prone to regular driving rains. Gable framing can be used to ensure that walls under gable ends also receive adequate protection. The use of straw bales as a roof insulation is not as widespread as its use as a wall insulation. The size and weight of straw bales are disadvantages that are difficult to overcome when using standard truss roof systems. Three different systems have been used to good advantage; bales insulating a flat ceiling, bales in box beam rafters, and bales in post and beam framing.

7.1 ROOF TYPES

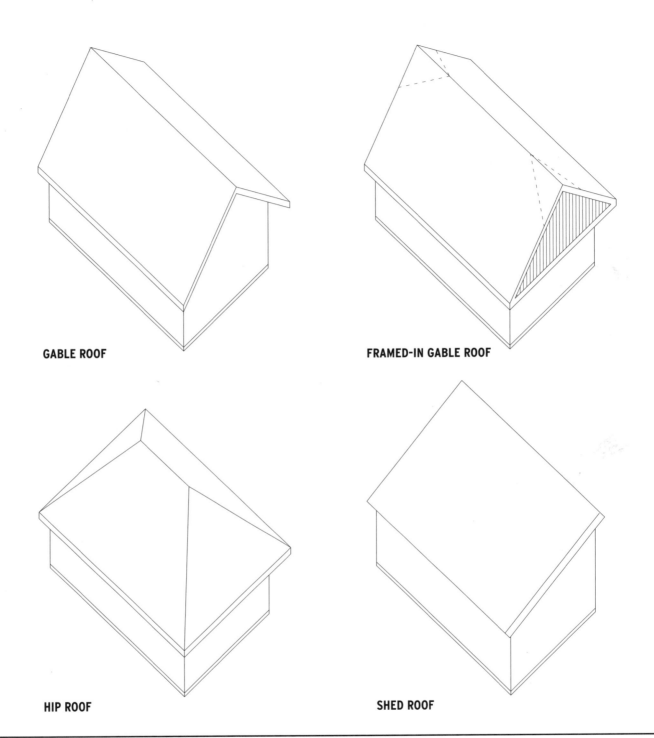

GABLE ROOF

FRAMED-IN GABLE ROOF

HIP ROOF

SHED ROOF

NOTES

1. Roofs should extend 18"–24" beyond wall
2. Gable and shed roof end walls not consistent in height
3. Framed-in gable and hip roofs have all walls of consistent height

NOT TO SCALE

7.2 ENGINEERED TRUSS ROOFS

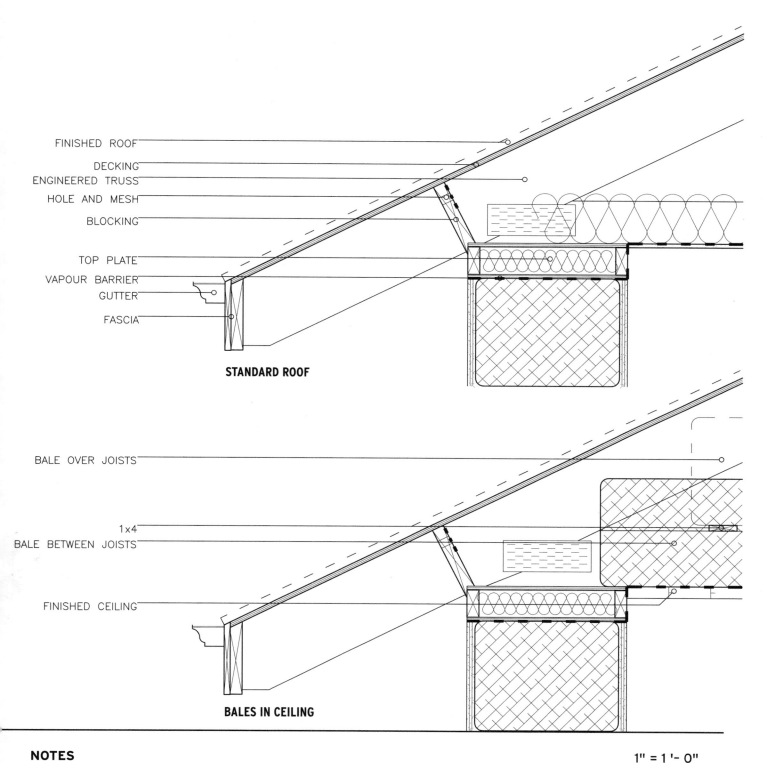

FINISHED ROOF

DECKING

ENGINEERED TRUSS

HOLE AND MESH

BLOCKING

TOP PLATE

VAPOUR BARRIER

GUTTER

FASCIA

STANDARD ROOF

BALE OVER JOISTS

1x4

BALE BETWEEN JOISTS

FINISHED CEILING

BALES IN CEILING

1" = 1'- 0"

NOTES

1. Bales in ceiling between joists require sufficient ceiling structure to support weight
2. Trusses spaced to accommodate bales between bottom cords
3. Bales in ceiling over bottom cord may require additional strapping for support (1x4 typ.)
4. Cover exposed vapour barrier with trim or cornice molding

7.3 BOX BEAM ROOFS

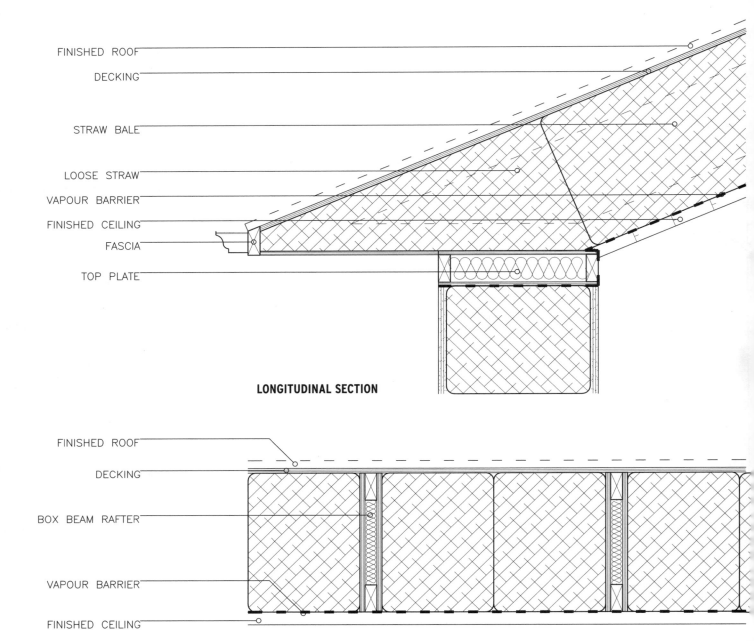

FINISHED ROOF

DECKING

STRAW BALE

LOOSE STRAW

VAPOUR BARRIER

FINISHED CEILING

FASCIA

TOP PLATE

LONGITUDINAL SECTION

FINISHED ROOF

DECKING

BOX BEAM RAFTER

VAPOUR BARRIER

FINISHED CEILING

TRANSVERSE SECTION

NOTES

1" = 1'- O"

1. Box beam to depth of bale width
2. Finished ceiling to support weight of bales
3. Roof overhang filled with loose straw
4. Cover exposed vapour barrier with trim or cornice molding

7.4 POST AND BEAM ROOFS

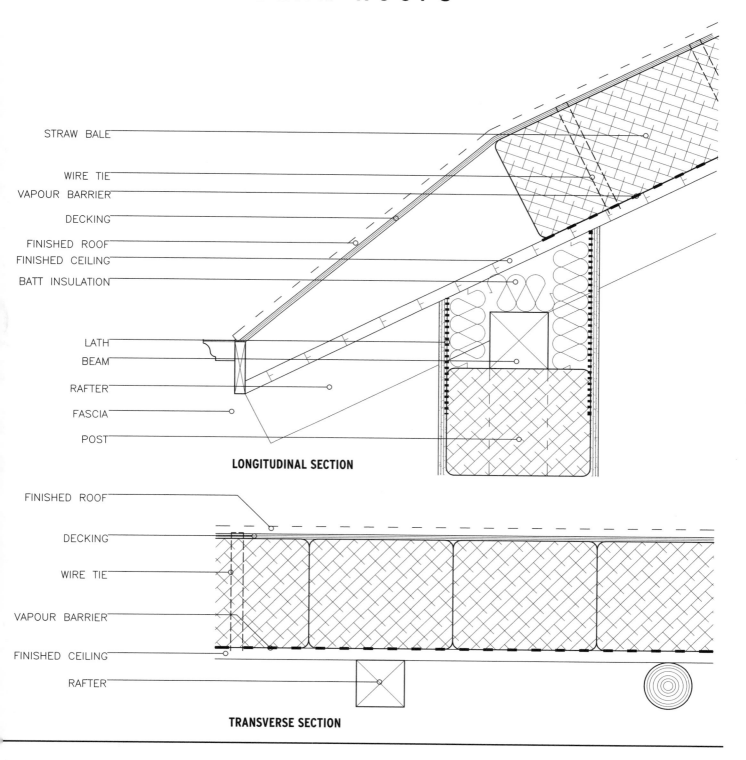

STRAW BALE

WIRE TIE
VAPOUR BARRIER

DECKING

FINISHED ROOF
FINISHED CEILING

BATT INSULATION

LATH
BEAM

RAFTER

FASCIA

POST

LONGITUDINAL SECTION

FINISHED ROOF

DECKING

WIRE TIE

VAPOUR BARRIER

FINISHED CEILING

RAFTER

TRANSVERSE SECTION

NOTES

1" = 1'- 0"

1. Posts and beams of square or round sections
2. Ceiling to support weight of bales
3. Roof decking tied to ceiling with wires through bales
4. Hollow between top course of wall and ceiling filled with batt insulation or equivalent

8.0 WALL-MOUNTING SYSTEMS, ELECTRICAL, PLUMBING

Wall-mounting Systems

Wall mounting systems are required to hang kitchen cabinets, pictures, and mount receptacles. Typical mounting systems use a 1x2 wooden stake embedded at least half way into the bale. Nails driven into the stake at an angle aid in preventing the stake from dislodging by acting like an arrow head. For kitchen cabinets, a 1x4 member is face nailed to the end of each stake. The 1x4 can either be attached to the finished wall surface or embedded in the plaster and covered over with the *finish coat*. For hanging pictures or other point loads, a nail can simply be nailed into the end of an embedded stake. After construction, picture hanging and other point loads can be handled by drilling into the plaster and using a wall plug. For post and beam structures, the posts can be used as mounting surfaces.

Electrical

To mount receptacles, a plywood board is sandwiched between two courses of bales and a notch is made in the upper course into which the receptacle is placed. Receptacles designed for R-2000 homes should be used with a plastic flange and neoprene seal against wires. If typical metal boxes are used, a 6 mil. polyethylene vapour barrier should be used behind the box. BX or romex cable is run through the joints in the bales.

Plumbing

Plumbing should not be embedded in the bale wall. Plumbing can, however, penetrate through the wall using a sleeve to avoid direct contact of the pipe with the bales. Plumbing can also be run along the floor–wall intersection as a base board element.

8.1 WALL MOUNTING SYSTEMS

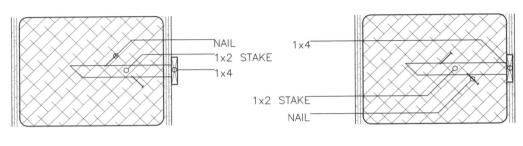

EXPOSED 1x4

NAIL
1x2 STAKE
1x4

RECESSED 1x4

1x4
1x2 STAKE
NAIL

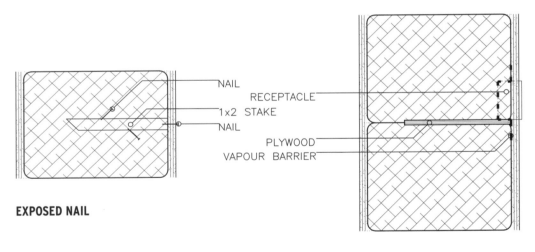

EXPOSED NAIL

NAIL
1x2 STAKE
NAIL

PLYWOOD PLATE

RECEPTACLE
PLYWOOD
VAPOUR BARRIER

NOTES

1. Stakes of 1x2s
2. Nails hammered at an angle to resist dislodging of stake
3. 1x4 rail at surface of wall or embedded in plaster skin
4. Nail in end of stake for hanging objects
5. Plywood plate between bales to support receptacles
6. Receptacles can be attached directly to posts

1" = 1'- 0"

8.2 ELECTRICAL & PLUMBING

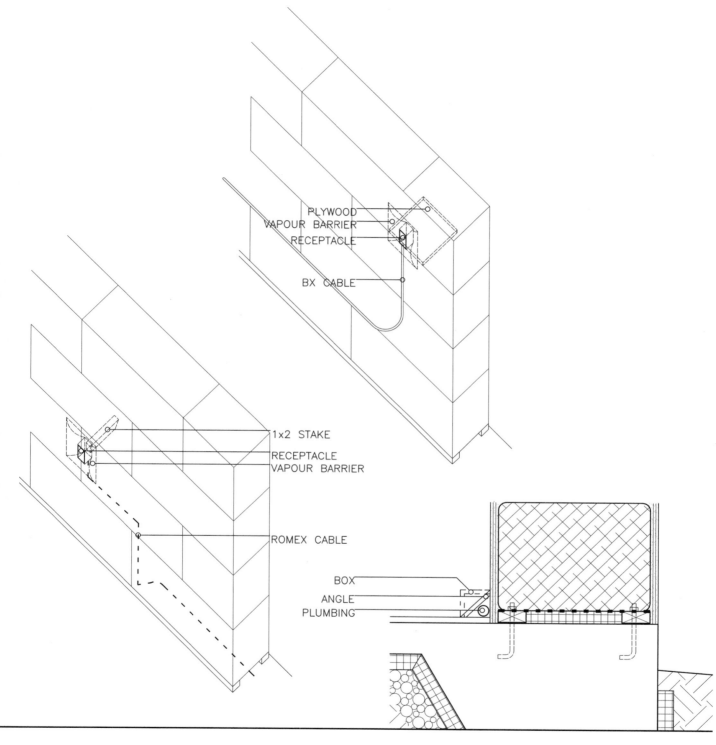

PLYWOOD
VAPOUR BARRIER
RECEPTACLE

BX CABLE

1x2 STAKE

RECEPTACLE
VAPOUR BARRIER

ROMEX CABLE

BOX
ANGLE
PLUMBING

NOT TO SCALE

NOTES

1. Electrical cables run between bales or at surface of bales
2. Vapour barrier required behind receptacles
3. Plumbing should not be in direct contact with bales
4. Baseboard runner can accommodate electrical or plumbing runs
5. Diamond lath around all receptacles

BIBLIOGRAPHY

EBNet Conference Proceedings, edited by Bruce King: CD-ROM available from Bruce King, Ecological Building Network, 209 Caledonia St., Sausalito, CA 94065 USA.

Edminster, Ann, *Investigation of Environmental Impacts of Straw Bale Construction*, 115 Angelita Ave., Pacifica, CA 94044 USA, 1995

Haggard, Ken and Scott Clark, Eds. *Straw Bale Construction Details: A Sourcebook*, The California Straw Building Association, Angels Camp, CA, 2000

King, Bruce, *Buildings of Earth and Straw: Structural Design for Rammed Earth and Straw Bale Architecture*, Ecological Design Press, Sausalito, CA, 1996

Lacinski, Paul and Michel Bergeron, *Serious Straw Bale: A Home Construction Guide for All Climates*, Chelsea Green Publishing Co., White River Junction, VT, 2000

Lerner, Kelly and Pamela Wadsworth Goode, *The Building Official's Guide to Straw Bale Construction*, The California Straw Bale Building Association, Angels Camp, CA, 2000

Magwood, Chris and Peter Mack, *Straw Bale Building: How to Plan, Design and Build with Straw*, New Society Publishers, Gabriola Island, BC, 2000

Meagan, Keely, of Earth Artisans, *Earth Plasters for Straw Bale Homes*, NM, 2000

Myhrman, Matts and S.O. MacDonald, *Build It with Bales: A Step-by-Step Guide to Straw-Bale Construction*, Out on Bale, Tucson, AZ, 1997

Steen, Athena and Bill, *The Beauty of Straw Bale Homes*, Chelsea Green Publishing Co., White River Junction, VT, 2000

Steen, Athena Swentzell and Bill and David Bainbridge, *The Straw Bale House*, Chelsea Green Publishing Co., White River Junction, VT, 2000

The Last Straw Journal: The Quarterly Journal of Strawbale and Natural Building, Networks Productions Inc., HC66 Box 119, Hillsboro, NM, 88042 USA

Thompson, Kim, *Straw Bale Construction: A Manual for Maritime Regions*, Straw House Herbals, Ship Harbour, NS, 1994

Straw Bale Building

How to Plan, Design and Build with Straw

Chris Magwood and Peter Mack

"The first major book published in the second wave of the straw-bale resurgence."
— *The Last Straw, the Journal of Straw Bale and Natural Building.*

Straw Bale Houses are easy to build, inexpensive, super energy efficient, environmentally friendly, attractive, and — best of all — can be designed to match the builder's personal space needs, aesthetics, and budget. Special focus on northern climates and building code compliance.
$24.95

The Art of Natural Building

Design, Construction, and Resources

Joseph E. Kennedy, Michael G. Smith and Catherine Wanek, Editors

"Highly recommended! Should be required reading for all students and builders."
— David A. Bainbridge, solar and straw bale pioneer, coauthor of *The Straw Bale House.*

The encyclopedia of natural building for non-professionals as well as architects and designers. From straw bale and cob, to recycled concrete and salvaged materials, this anthology from leaders in the field focuses on the practical and aesthetic concerns of ecological building.
$26.95

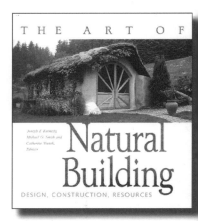

If you have enjoyed *Straw Bale Details* you might also enjoy other

BOOKS TO BUILD A NEW SOCIETY

Our books provide positive solutions for people who want to make a difference. We specialise in:

**Sustainable Living • Ecological Design and Planning • Natural Building & Appropriate Technology
New Forestry • Environment and Justice • Conscientious Commerce • Progressive Leadership
Educational and Parenting Resources • Resistance and Community • Nonviolence**

For a full list of NSP's titles, please call 1-800-567-6772 or check out our web site at:

www.newsociety.com

NEW SOCIETY PUBLISHERS